Collins

The Shanghai Maths Project

For the English National Curriculum

一课一练

Year 5 Learning

Collins

William Collins' dream of knowledge for all began with the publication of his first book in 1819.

A self-educated mill worker, he not only enriched millions of lives, but also founded a flourishing publishing house. Today, staying true to this spirit, Collins books are packed with inspiration, innovation and practical expertise. They place you at the centre of a world of possibility and give you exactly what you need to explore it.

Collins. Freedom to teach.

Published by Collins
An imprint of HarperCollins*Publishers*
The News Building
1 London Bridge Street
London
SE1 9GF

Browse the complete Collins catalogue at
www.collins.co.uk

Learning Books Series Editor: Amanda Simpson

Practice Books Series Editor: Professor Lianghuo Fan

Authors: Sarah Eaton, Linda Glithro, Paul Hodge, Jane Jones, Steph King, Richard Perring, Paul Wrangles

British Library Cataloguing in Publication Data

A catalogue record for this publication is available from the British Library.

MIX
Paper from responsible sources
FSC www.fsc.org FSC™ C007454

This book is produced from independently certified FSC paper to ensure responsible forest management.

For more information visit: **www.harpercollins.co.uk/green**

Publishing Manager: Fiona McGlade and Lizzie Catford
In-house Editor: Mike Appleton
In-house Editorial Assistant: August Stevens
Project Manager: Karen Williams
Copy Editor: Karen Williams
Proofreader: Catherine Dakin
Cover design: Kevin Robbins and East China Normal University Press Ltd
Cover artwork: Daniela Geremia
Internal design: Amparo Barrera
Typesetting: Ken Vail Graphic Design Ltd
Illustrations: QBS
Production: Sarah Burke

Printed and bound in Latvia

Photo acknowledgements
The publishers wish to thank the following for permission to reproduce photographs. Every effort has been made to trace copyright holders and to obtain their permission for the use of copyright materials. The publishers will gladly receive any information enabling them to rectify any error or omission at the first opportunity.

(t = top, c = centre, b = bottom, r = right, l = left)
p43 cb Plan-B/Shutterstock

Pure decimal

(read as 'zero point two seven three')

0.273

Whole number part is zero.

Decimal part is not zero.

Mixed decimal

(read as 'seven point one zero three')

7.103

Whole number part is not zero.

Decimal part is not zero.

Properties of decimals

Zeros can be written or removed from the end of the decimal part of a number without changing its value.

0.4 and 0.40 have the same value.

0.04 and 0.040 have the same value.

3.04 and 3.040 have the same value.

0.04 and 0.40 do **not** have the same value.

Representations

Place value chart (showing 0.75)

100	10	1	.	0.1	0.01
		0	.	7	5

100	10	1	.	0.1	0.01
			.	●●●● ●●●	●●● ●●

Gattegno chart (showing 0.75)

Ten thousands	10 000	20 000	30 000	40 000	50 000	60 000	70 000	80 000	90 000
Thousands	1000	2000	3000	4000	5000	6000	7000	8000	9000
Hundreds	100	200	300	400	500	600	700	800	900
Tens	10	20	30	40	50	60	70	80	90
Ones	1	2	3	4	5	6	7	8	9
Tenths	0.1	0.2	0.3	0.4	0.5	0.6	(0.7)	0.8	0.9
Hundredths	0.01	0.02	0.03	0.04	(0.05)	0.06	0.07	0.08	0.09
Thousandths	0.001	0.002	0.003	0.004	0.005	0.006	0.007	0.008	0.009

Place value counters (showing 0.75)

0.1 0.1 0.1 0.1   0.01 0.01

0.1 0.1 0.1   0.01 0.01 0.01

Place value slider (showing 0.75)

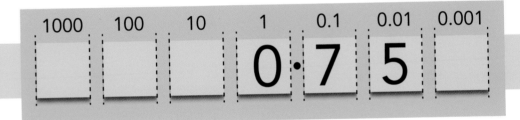

1000	100	10	1	0.1	0.01	0.001

0·75

Number line (showing 0.75)

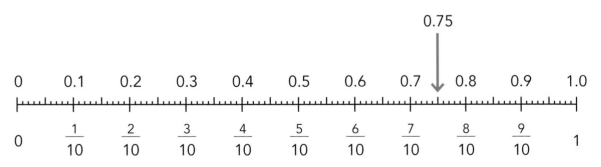

0.75

0   0.1   0.2   0.3   0.4   0.5   0.6   0.7   0.8   0.9   1.0

0 $\frac{1}{10}$ $\frac{2}{10}$ $\frac{3}{10}$ $\frac{4}{10}$ $\frac{5}{10}$ $\frac{6}{10}$ $\frac{7}{10}$ $\frac{8}{10}$ $\frac{9}{10}$ 1

Fractions and decimals

$$\frac{1}{10} = 0.1 \qquad \text{so} \qquad \frac{2}{10} = 0.2$$

$$\frac{1}{100} = 0.01 \qquad \text{so} \qquad \frac{7}{100} = 0.07$$

$$\frac{1}{1000} = 0.001 \qquad \text{so} \qquad \frac{3}{1000} = 0.003$$

A decimal number can be written as the sum of fractions with denominator 10, 100, 1000, …

$$\frac{2}{10} + \frac{7}{100} + \frac{3}{1000} = 0.273$$

Order of operations

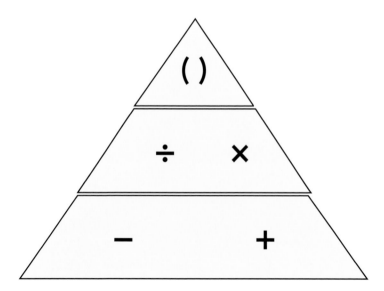

Laws of operations

The associative law

$a \times (b \times c) = (a \times b) \times c$

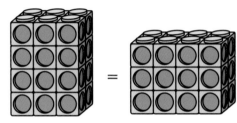

$4 \times (2 \times 3) = (4 \times 2) \times 3$

The distributive law

$a \times (b + c) = (a \times b) + (a \times c)$

$5 \times (3 + 6) \quad = \quad (5 \times 3) + (5 \times 6)$

This array represents how $23 \times 45 + 23 \times 25$ can be visualised and calculated.

	45	25
23		

Equivalence of measures

Kilograms and grams

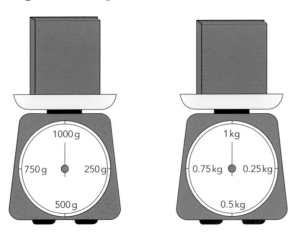

1 kilogram = 1000 grams

Litres and millilitres

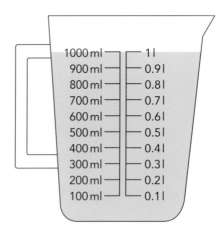

1 litre = 1000 millilitres.

Converting between measurements

Step 1: Find what one unit is worth.

Step 2: If you are converting from a smaller unit to a larger unit, <u>divide</u>. If you are converting from a larger unit to a smaller unit, <u>multiply</u>.

For example: 5230 grams = ? kilograms

Step 1: There are 1000 g in 1 kg.

Step 2: Converting from a smaller unit (g) to a larger unit (kg).

So: 5320 ÷ 1000 = 5.32 kg

Multiplying and dividing by 10, 100 or 1000

Shift the digits to the left (when multiplying) or right (when dividing) the number of zeros in the multiplier/divisor. For example:

6.3 × 1000

1000	100	10	1	.	0.1	0.01
			6		3	
6	3	0	0			

6.3 × 1000 = 6300

514 ÷ 1000

100	10	1	.	0.1	0.01	0.001
5	1	4				
		0	.	5	1	4

514 ÷ 1000 = 0.514

Reading and writing very large numbers

Billions			Millions			Thousands			Ones		
Hundreds	Tens	Ones	Hundreds	Tens	Ones	Hundreds	Tens	Ones	Hundreds	Tens	Ones

Numbers are separated into groups of three, beginning from the **right, 3210654987 becomes 3 210 654 987**.

Numbers are then read from the **left**, beginning with the largest group, three billion, two hundred and ten million, six hundred and fifty-four thousand, nine hundred and eighty-seven.

Rounding very large numbers

Rounding numbers makes them simpler and easier to use. Using rounded numbers to estimate an answer is useful and is often used for checking.

To round a number, decide which is the last digit to keep. Increase it by 1 if the next digit is 5 or greater and all digits to its right become zero. Leave it unchanged if the digit is less than 5 and all digits to its right become zero. (By convention, the mid-point (5) is rounded to the next whole 100, 1000, ...)

Take the number **5 463 728** (five million, four hundred and sixty-three thousand seven hundred and twenty-eight).

Rounded to the nearest 100, it becomes **5 463 700**

Rounded to the nearest 1000, it becomes **5 464 000**

Rounded to the nearest 10 000, it becomes **5 460 000**

Rounded to the nearest 100 000, it becomes **5 500 000**

Rounded to the nearest 1 000 000, it becomes **5 000 000**

Sometimes you may be asked to round a number **up** or **down** rather than to the nearest 100, 1000, ..., so read the instructions carefully!

For example, to round **5 463 728 up** to a whole number of millions, it becomes **6 000 000**.

Speed, time and distance

Sally is the fastest as she took the least time to run 800 m.

Asha is the slowest as she took the most time to run the same distance.

We can calculate the actual speed of each runner by dividing the distance by the time taken.

Jamil took 4 minutes to run 800 m.

His speed can be found by dividing 800 by 4. His speed was 200 metres per minute (m/min).

8 0 0 M R A C E

1	SALLY	2	MINUTES
2	JAMIL	4	MINUTES
3	ASHA	5	MINUTES

I ran at a speed of 100 m/min but I have forgotten my time.

The time taken can be calculated by dividing the distance travelled by the speed.

I ran at a speed of 120 m/min for 6 minutes but I didn't finish the race.

The distance travelled can be calculated by multiplying the speed by the time taken.

Equivalent divisions

The arrays show 12 ones divided by 3 ones, 12 tens divided by 3 tens and 12 tens divided by 3 ones.

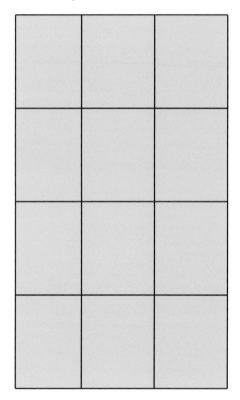

$$12 \div 3 = 4$$

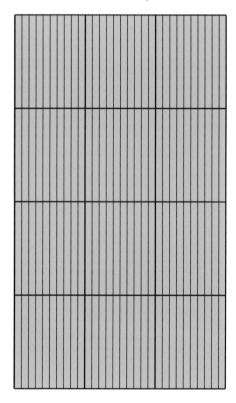

$$120 \div 30 = 4$$

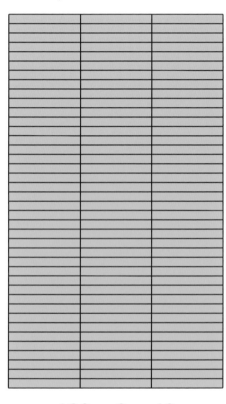

$$120 \div 3 = 40$$

The divisions 12 ÷ 3 and 120 ÷ 30 are equivalent because they both result in the same quotient.

The dividend and the divisor in 12 ÷ 3 are both scaled by ten in 120 ÷ 30 so the relationship between them stays the same.

The quotient for 120 ÷ 3 is ten times greater than for 12 ÷ 3 because only the dividend is 10 times larger.

Estimating quotients

$$345 \div 40$$

I estimated the quotient as 8 using $340 \div 40$ and the equivalent division $34 \div 4$.

I think the quotient will be just right because 8×40 is 320. The remainder is not large enough to make another group of 40.

The children make the same estimate for the division $345 \div 46$.

This time, the quotient 8 is too large because $8 \times 46 = 368$. The quotient must be 7.

When the divisor (46) is closer to the next ten (50), it is likely that the quotient will need to be adjusted.

We can also use the division $340 \div 50$ help make an estimate.

We can also estimate the size of quotients by comparing the dividend and the divisor.

$345 \div 46$ There are 34 tens to be divided by 46. The quotient must be a one-digit number because 46×10 is 460. We can see that $34 < 46$.

$345 \div 33$ There are 34 tens to be divided by 33. The quotient must be a two-digit number because 33×10 is 330. We can see that $34 > 33$.

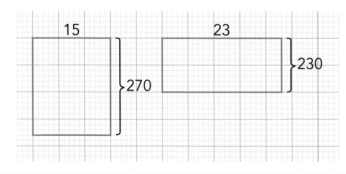

For the divisions $270 \div 15$ and $230 \div 23$, we can see that the quotients are also 2-digit numbers.

Mental calculations

We can use what we know about the relationship between multiplication and division to help solve calculations mentally.

a)

350 ÷ 55	When 35 ÷ 5, the quotient is 7.
	7 × 55 = 385
	The quotient is too big.
	Change the quotient to 6.
	6 × 55 = 330
	The remainder is 20. The quotient of 6 is correct.

Arrays can be used to represent divisions.

Here we can see that the quotient 7 is too big.

b) We can also use partitioning to find parts of the quotient before recombining.

272 ÷ 34

34 × 9 = 306 (quotient is too big)

34 × 6 = 204 (quotient is too small because more groups of 34 can be made)

First 204 ÷ 34 = 6

Then 68 ÷ 34 = 2

Therefore, 272 ÷ 34 = 6 + 2

The array shows the dividend 272 partitioned into 204 and 68.

Column method

The base 10 blocks represent the division 185 ÷ 50.

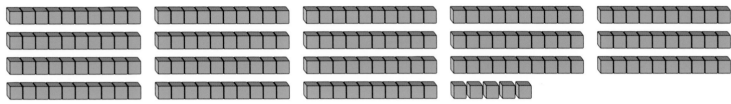

We can record this in the column method.

```
        0 0 3   remainder 35
5 0 ) 1 8 5
        1 5 0   (subtract 3 × 50)
            3 5
```

Nothing can be entered in the hundreds or tens column of the answer because 185 is less than 10 times 50. The highest place value in the quotient is the ones.

We can use the same method to divide much larger numbers.

```
          5 6
4 8 ) 2 6 8 8
      2 4 0
          2 8 8
          2 8 8   (subtract 6 × 48)
          0 0 0
```

The divisor 48 is closer to 50 so the quotient will be closer to the estimate 2600 ÷ 50 than it is to 2600 ÷ 40.

I know that the quotient will be a 2-digit number because the first two digits in the 4-digit dividend is less than the divisor. The highest place value in the quotient is the tens.

Working with the four operations

Brackets can be used to show what needs to be carried out first in a calculation.

$(30 \times 6) - (24 \div 6) = 180 - 4$
$= 176$

Multiplication and division must also be carried out before addition or subtraction.

$6 \times 30 + 24 \div 6 = 180 + 4$
$= 184$

When calculations involve only multiplication and division, we can re-order them to make the calculation easier.

4200 (× 11) (÷ 60) can be re-ordered as 4200 (÷ 60) (× 11) so the division fact $42 \div 6$ can be used.

250 (× 4) (÷ 20) is best carried out in the order shown because $250 \times 4 = 1000$ and can be easily divided by 20.

Comparing fractions

$\frac{1}{9}$	$\frac{1}{9}$	$\frac{1}{9}$	$\frac{1}{9}$	$\frac{1}{9}$	$\frac{1}{9}$	$\frac{1}{9}$	$\frac{1}{9}$	$\frac{1}{9}$

$\frac{1}{3}$	$\frac{1}{3}$	$\frac{1}{3}$

$\frac{1}{9}$	$\frac{1}{9}$	$\frac{1}{9}$	$\frac{1}{9}$	$\frac{1}{9}$	$\frac{1}{9}$	$\frac{1}{9}$	$\frac{1}{9}$	$\frac{1}{9}$

$\frac{1}{6}$	$\frac{1}{6}$	$\frac{1}{6}$	$\frac{1}{6}$	$\frac{1}{6}$	$\frac{1}{6}$

$\frac{1}{9}$	$\frac{1}{9}$	$\frac{1}{9}$	$\frac{1}{9}$	$\frac{1}{9}$	$\frac{1}{9}$	$\frac{1}{9}$	$\frac{1}{9}$	$\frac{1}{9}$

$\frac{2}{9} < \frac{7}{9}$

For fractions that have the same denominator, smaller fractions have smaller numerators.

$\frac{1}{3} > \frac{1}{6} > \frac{1}{9}$

For a fraction with a numerator of 1, smaller fractions have greater denominators.

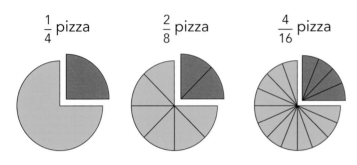

$\frac{1}{4}$ pizza $\frac{2}{8}$ pizza $\frac{4}{16}$ pizza

$\frac{1}{5}$	$\frac{1}{5}$	$\frac{1}{5}$	$\frac{1}{5}$	$\frac{1}{5}$

$\frac{1}{7}$	$\frac{1}{7}$	$\frac{1}{7}$	$\frac{1}{7}$	$\frac{1}{7}$	$\frac{1}{7}$	$\frac{1}{7}$

$\frac{1}{9}$	$\frac{1}{9}$	$\frac{1}{9}$	$\frac{1}{9}$	$\frac{1}{9}$	$\frac{1}{9}$	$\frac{1}{9}$	$\frac{1}{9}$	$\frac{1}{9}$

$\frac{1}{4} = \frac{2}{8} = \frac{4}{16}$

For equivalent fractions, the numerator and the denominator are multiplied or divided by the same amount.

$\frac{3}{9} < \frac{3}{7} < \frac{3}{5}$

For a fraction with the same numerator, smaller fractions have greater denominators.

Proper fractions, improper fractions and mixed numbers

Proper fractions	Improper fractions	Mixed numbers
$\frac{3}{5}$	$\frac{4}{3}$	$1\frac{1}{3}$
A proper fraction is a fraction where the numerator is less than the denominator.	An improper fraction is a fraction where the numerator is greater than the denominator.	A mixed number is a whole number and a fraction combined into one number.
Example: $\frac{5}{9}, \frac{7}{12}, \frac{76}{85}, \frac{132}{135}$	Example: $\frac{7}{5}, \frac{15}{14}, \frac{35}{27}, \frac{189}{135}$	Example: $2\frac{4}{5}, 5\frac{7}{9}, 12\frac{1}{2}, 24\frac{5}{7}$

Adding and subtracting fractions with related denominators

$$\frac{7}{6} + \frac{9}{12}$$

$$= \frac{14}{12} + \frac{9}{12}$$

$$= \frac{14 + 9}{12}$$

$$= \frac{23}{12} = 1\frac{11}{12}$$

← When adding or subtracting fractions with related denominators, one fraction needs to be converted so both fractions have the same denominator. →

$$\frac{13}{16} - \frac{3}{8}$$

$$= \frac{13}{16} - \frac{6}{16}$$

$$= \frac{13 - 6}{16}$$

$$= \frac{7}{16}$$

Multiplying and dividing by 10, 100, 1000

The digits move across the decimal point

They move one place to the left for every 10 that you multiply by and one place to the right for every 10 that you divide by.

100	10	1	.	0.1	0.01
		3	.	1	
	3	1	.		

3.1 × **10** = 31

All of the digits move **one** place left across the decimal point.

Remember to fill any gaps to the left with zeros.

52 ÷ **100** = 0.52

All of the digits move **two** places right across the decimal point.

100	10	1	.	0.1	0.01
	5	2	.		
		0	.	5	2

It can be helpful to imagine a place value slider when multiplying and dividing by 10, 100 or 1000.

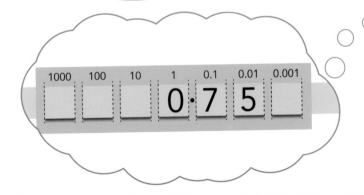

1000	100	10	1	0.1	0.01	0.001
			0 . 7	5		

Units

Kilo means 1000 of a unit, so a kilogram is 1000 grams and a kilometre is 1000 metres.

Centi means $\frac{1}{100}$ of a unit, so a centimetre is $\frac{1}{100}$ of a metre and a centilitre is $\frac{1}{100}$ of a litre.

Milli means $\frac{1}{1000}$ of a unit, so a millimetre is $\frac{1}{1000}$ of a metre and a milligram is $\frac{1}{1000}$ of a gram.

> You will need to think about which is the best unit for each question.

Carrying out calculations with units is easier if all measurements are written using the same units.
- Convert all of the weights to grams or kilograms before calculating.
- Convert all of the lengths to millimetres, centimetres, metres or kilometres before calculating.

Adding and subtracting decimals

Adding and subtracting decimals using the column method follows all of the same rules as adding and subtracting whole numbers. **Just remember to line up the decimal points one above the other!**

1.369 + 7.58 = …

```
  1 . 3 6 9
+ 7 . 5 8
---------
  8 . 9 4 9
```

The decimal points are lined up

3.142 − 1.071 = …

```
  3 . ⁰1¹4 2
− 1 . 0 7 1
---------
  2 . 0 7 1
```

The decimal points are lined up

Positive and negative numbers

On a number line, positive numbers are to the right of the origin and negative numbers are to the left. The numbers continue in both directions indefinitely. Each side of the origin mirrors the other.

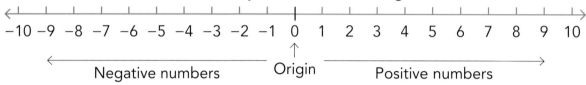

← Getting smaller	Getting bigger →
Any number to the left of a number is smaller, so –6 is smaller than 4, –6 < 4 and –8 is smaller than –5, –8 < –5.	Any number to the right of a number is bigger, so –2 is bigger than –10, –2 > –10 and 3 is bigger than –1, 3 > –1.

–5°C is 5° below 0°C	A bank balance of –£5 means you owe £5	Depths below sea-level are shown as negative

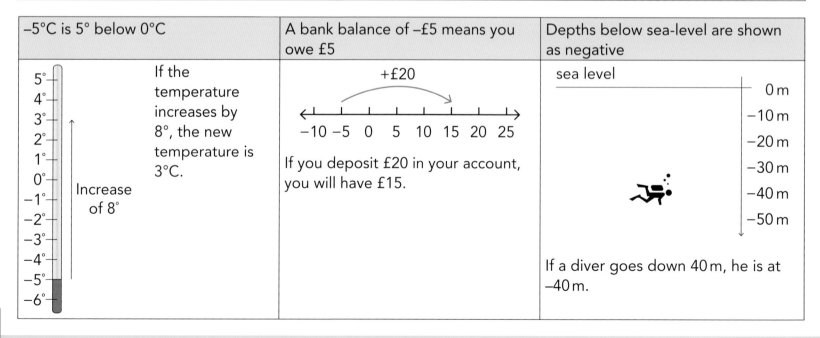

If the temperature increases by 8°, the new temperature is 3°C.

Increase of 8°

+£20

If you deposit £20 in your account, you will have £15.

sea level

If a diver goes down 40 m, he is at –40 m.

Parts of a circle

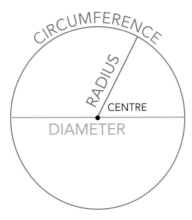

diameter = 2 × radius
$d = 2 \times r$

Naming angles

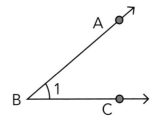

You can name angles in different ways:

- the vertex ($\angle B$)
- refer to the lines ($\angle ABC$ or $\angle CBA$) (the vertex is the middle letter)
- a number ($\angle 1$).

Vertically opposite angles

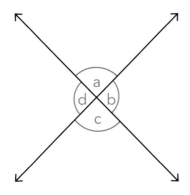

Vertically opposite angles are equal.
$\angle a$ and $\angle c$ are vertically opposite $\angle a = \angle c$
$\angle b$ and $\angle d$ are vertically opposite $\angle b = \angle d$

The measurement of angles

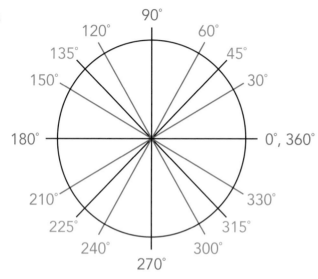

Angles can be measured in degrees.

The symbol for degrees is a small circle °.

For example, an angle of 30 degrees is written as 30°.

One degree is equivalent to $\frac{1}{360}$ of a full rotation and 360 degrees means one full rotation.

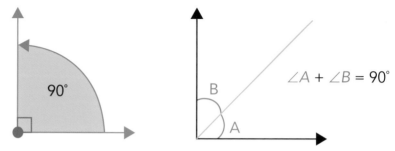

A right angle is 90 degrees, or 90°.

Angles that combine to make a right angle will have a sum of 90°.

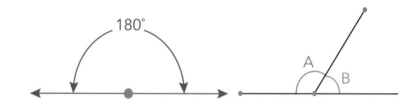

A straight angle contains two right angles, so it is 180°.

Angles on a straight line that combine to make a straight angle will have a sum of 180°.

$\angle A + \angle B = 180°$

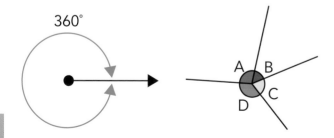

A full angle contains 4 right angles, so it is 360°. Angles around a point that combine to make a full angle will have a sum of 360°.

$\angle A + \angle B + \angle C + \angle D = 360°$

Cubes and cuboids

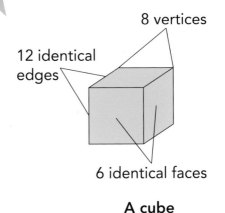

8 vertices

12 identical edges

6 identical faces

A cube

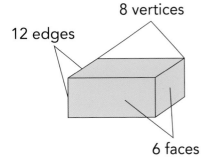

8 vertices

12 edges

6 faces

A cuboid

Cubic centimetres and cubic metres

(Not drawn to scale.)

1 cubic centimetre (1 cm^3)

1 cm
1 cm
1 cm

1 cubic metre (1 m^3)

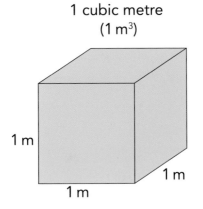

1 m
1 m
1 m

Finding the volume of a shape made from centimetre cubes

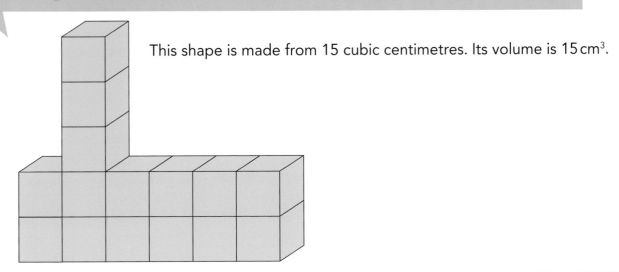

This shape is made from 15 cubic centimetres. Its volume is 15 cm^3.

Finding the volume of a cube or cuboid

$l \times w \times h$ = volume

where l = length, w = width and h = height

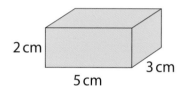 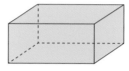

2 cm

5 cm

3 cm

l = 5 cm, w = 3 cm, h = 2 cm

$5 \times 3 \times 2 = 30$

volume = 30 cm^3

Volume is measured in cubic units, for example cubic centimetres (cm^3) or cubic metres (m^3).

Volume and capacity

1 cm^3 = 1 ml

1000 cm^3 = 1000 ml = 1 l

Imperial and metric measurements

1 inch ≈ 2.5 cm

1 pint ≈ 570 ml or 0.57 l

1 pound ≈ 450 g or 0.45 kg

2.2 pounds ≈ 1 kg

Types of number

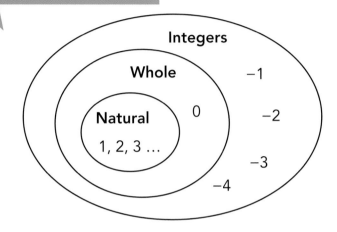

Divisibility

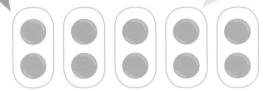

10 is divisible by 5

$10 \div 5 = 2$

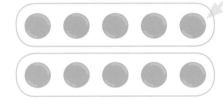

10 is divisible by 2

$10 \div 2 = 5$

28 is divisible by 4

28	÷	4	=	7
dividend		divisor		quotient
28	÷	7	=	4
dividend		divisor		quotient

Factors and multiples

$$5 \times 4 = 20$$

→ multiple of 4

→ multiple of 5

factor of 20   factor of 20

Factors of 16

1 × 16 or 16 × 1	2 × 8 or 8 × 2	4 × 4

The product for each multiplication is 16. The integers 1, 16, 2, 8 and 4 are all factors of 16.

Multiples of 12

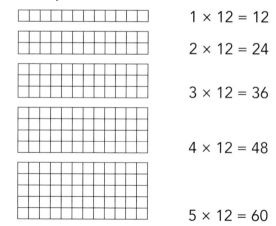

1 × 12 = 12

2 × 12 = 24

3 × 12 = 36

4 × 12 = 48

5 × 12 = 60

Square numbers

A square number is when an integer is multiplied by itself.

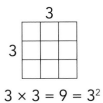

$3 \times 3 = 9 = 3^2$

Cube numbers

When a value is multiplied by itself three times, the product is called a cube number.

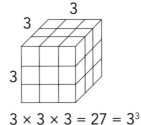

$3 \times 3 \times 3 = 27 = 3^3$

Adding square numbers

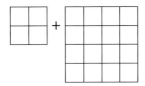

$2 \times 2 + 4 \times 4 = 2^2 + 4^2$

Adding consecutive numbers

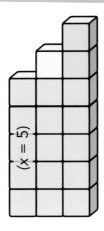

(x = 5)

$x + (x + 1) + (x + 2)$ or $3x + 3$

- $x = 5$ or $5 = x$
- there are 3 lots of x here
- the second number is the same as x, with one extra cube, so $x + 1$
- the third number is the same as x but with 2 extra cubes, so $x + 2$
- 3 lots of x add 3 extra cubes gives the total of the 3 numbers

Prime numbers and composite numbers

A **prime number** is a number (greater than 1) that cannot be divided by any number except itself and 1.

7 is prime, its only factors are 7 and 1.

A **composite number** is a number that can be divided by itself and 1, and by other numbers.

10 is composite.

Its factors are 1, 2, 5 and 10.

All whole numbers greater than 1 are either composite or prime.

The prime numbers up to 100 are **2, 3, 5, 7, 11, 13, 17, 19, 23, 29, 31, 37, 41, 43, 47, 53, 59, 61, 67, 71, 73, 79, 83, 89, 97**.

2 is the only even prime number because every other even number can be divided into equal groups.

Prime factorisation

A **prime factor** is a factor that is a prime number.

Every composite number has a unique set of **prime factors** that give the number when they are multiplied.

$$30 = 2 \times 3 \times 5 \qquad 31 = 31 \qquad 32 = 2 \times 2 \times 2 \times 2 \times 2$$

$$33 = 3 \times 11 \qquad 34 = 2 \times 17 \qquad 35 = 5 \times 7$$

Factors and prime factors

The factors of **30** are 1, 2, 3, 5, 6, 10, 15, 30 giving factor pairs $1 \times 30, 2 \times 15, 3 \times 10, 5 \times 6$.

The prime factors are 2, 3 and 5 because $2 \times 3 \times 5 = 30$.

The factors of **100** are 1, 2, 4, 5, 10, 20, 25, 50, 100 giving factor pairs $1 \times 100, 2 \times 50, 4 \times 25, 5 \times 20, 10 \times 10$.

The prime factors are 2, 2, 5, 5 because $2 \times 2 \times 5 \times 5 = 100$.

2-D shape: A flat shape that can be shown on screen or on paper. It does not have any thickness so cannot be handled.

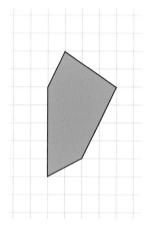

3-D shape: A shape with three dimensions. A cuboid is a 3-D shape.

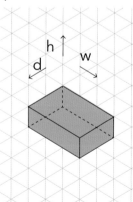

acute angle: An angle that is smaller than a right angle. It is greater than 0 degrees and less than 90 degrees.

addend: a number being added, or added to, in an addition calculation: addend + addend = sum

$$45 + 30 = 75$$
$$\uparrow \quad \uparrow$$
$$\text{addend}$$

addition: An operation in which two or more numbers are combined or one number is increased by another. The symbol for addition is +.

$$247 + 152 = 399$$

angle: the amount of turn between two straight lines that meet at a point

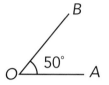

To move line *OA* onto *OB* requires a turn of 50°.

arc: a segment of the circumference of a circle

area: the amount of space occupied by a 2-D object

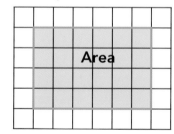

as many: 15 times as many is the same as multiplying by 15

associative law: a rule in mathematics that states factors may be associated in any way without changing the outcome, for example $(2 \times 4) \times 3 = 24$ and $2 \times (4 \times 3) = 24$

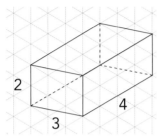

benchmark angle: angles such as 30 degrees, 45 degrees, 60 degrees, 90 degrees which can be used as references

brackets: symbols used to contain parts of a calculation, to clarify and indicate the order in which parts of the calculation should be carried out

$$2 \times (45 - 10) = 70$$

calculation: The act of calculating. Also, the written representation of the calculation using numbers and symbols can be called 'a calculation'.

capacity: the amount of material (gas, liquid or solid) that a container can contain

centimetre: a unit of measure equivalent to 10 millimetres

centre: the point equidistant from all points on the edge of a circle

check: work something out again to make sure it makes sense and is correct

circumference: the edge of a circle

column: a vertical arrangement of numbers or objects, or a shaded vertical block

commutative law: a rule in mathematics which states that in addition and multiplication, the numbers can be swapped around without changing the answer

$2 \times 50 = 100$	$50 \times 2 = 100$
$a + b = b + a$	$a \times b = b \times a$

composite number: a whole number that can be divided equally by numbers other than 1 or itself

consecutive (numbers): numbers that follow each other in order without gaps

6, 7, 8 are consecutive numbers

convert: to change into a different, equivalent form

convert (units): to change a measurement expressed in one unit into an equivalent measurement expressed in another unit

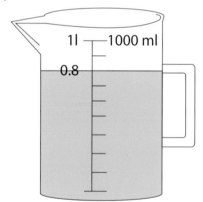

$0.8 \times 1000 = 800$ and $0.8 l = 800 ml$.

cube: a 3-D shape with 6 identical square faces, 12 identical edges and 8 vertices

cube number: A number formed by multiplying a digit by itself three times. For example, $4 \times 4 \times 4 = 64$. 64 is a cube number.

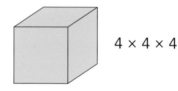

$4 \times 4 \times 4$

cubic centimetre: A cube with length 1 cm, width 1 cm and height 1 cm. Cubic centimetres are a unit of measure used to measure volume and capacity.

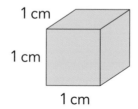

1 cm
1 cm
1 cm

cubic metre: A cube with length 1 m, width 1 m and height 1 m. Cubic metres are a unit of measure used to measure volume and capacity.

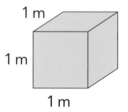

1 m
1 m
1 m

cuboid: a 3-D shape with 6 rectangular faces (or 4 rectangular and 2 square faces), 12 edges and 8 vertices

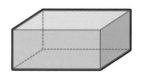

decimal number: a number between two whole numbers, in which tenths, hundredths, thousandths and so on, are represented by digits in columns to the right of the decimal point

decimal places: the number of digits written in the decimal part of a number

467.23 is an example of a number written to 2 decimal places

decimal point: The symbol (a dot) used to separate the whole number part from the decimal part. In printed materials, like books, it is usually shown on the line like a full stop. When handwritten, it is written above the line, half way up.

      1604.78   1604·78

decimal point

degree: A unit of measure that can be used to describe the amount of turn. One degree is equivalent to a $\frac{1}{360}$th turn of a circle.
The symbol for degree is °.

denominator: The number of equal parts an object, quantity or number has been divided into. When a fraction is written, the denominator is the bottom number.

$$\frac{7}{12} \longleftarrow \text{denominator}$$

diagonal: a straight line that goes from one corner to an opposite corner

diameter: any line that joins two points on a circle and passes through the centre

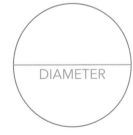

DIAMETER

difference: the result of a subtraction calculation: minuend – subtrahend = difference

distance: how far an object travels; the length between two points

distributive law: A rule that means that adding a group of numbers together and then multiplying the total by something is the same as multiplying them separately and then adding them together. It is also true for division.

$$a \times (b + c) = a \times b + a \times c$$
$$\text{and}$$
$$a \times (b - c) = a \times b - a \times c$$

divide: to separate into equal parts

dividend: the whole quantity before it is divided

In $5400 \div 30 = 180$, the dividend is 5400

divisibility test: A quick test to see if one number is divisible by another. For example, even numbers are always divisible by 2; numbers ending in 5 or 0 are always divisible by 5.

divisible: a dividend is divisible by a number only when the quotient is an integer

$$\frac{12}{3} = 4.$$
Therefore, 12 is divisible by 3

---

division: An operation where something is divided, that is, a number is split into equal parts or groups. A symbol for division is ÷

$$200 \div 50 = 4$$

---

divisor: the number by which another number is divided

$$12 \div 4 = 3$$
The divisor is 4

---

double: twice as many, or twice as much

edge(s): a line that joins two vertices on a 3-D shape

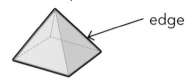
edge

---

efficient strategy: a way of working out an answer quickly and accurately

---

equal groups: groups of the same value

---

equal parts: parts of the same size

---

equals: has the same value; the symbol for equals is =

---

equivalent: equal in quantity, size or value; have the same effect

---

estimate: to decide approximately what the answer will be

expression: when numbers, symbols and operations are connected to communicate relationships, for example $\frac{64}{8} + 40$ is an expression that tells us that 64 is divided by 8 and the result is added to 40

$$\frac{64}{8} + 40$$

---

exterior angle: an angle created by extending the side of a polygon angle on the outside of a polygon.

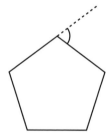

---

face(s): a flat surface of a 3-D shape, bounded by edges

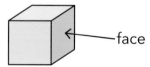
face

factor: a number that divides exactly into a another number, for example in 4 × 5 = 20, 4 and 5 are both factors of 20

$$4 \times 5 = 20$$

Factors of 20

foot: an imperial unit of measure used to measure length (1 foot = 12 inches)

full angle: an angle of 360 degrees, equal to a full rotation

360 degrees

function machine: An imaginary machine that changes the numbers that are fed into it in particular ways. A diagram of a function machine represents a set of operations which can be applied to any number.

input

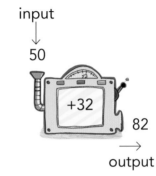

50

+32

82

output

gram: A unit of measure used to measure mass (1 gram = 0.001 kilograms).

A paper clip has a mass of approximately 1 gram.

heavier: more heavy; has more mass

heptagon: a 2-D shape with 7 straight sides and 7 vertices

hexagon: a 2-D shape with 6 straight sides and 6 vertices

horizontal: at a right angle to the vertical; going side-to-side

hundredth: when 1 whole is split into 100 equal parts, each part is a hundredth

imperial: A system of measurement that uses units such as inches and feet to measure distance, pounds to measure mass and pints to measure liquid volume. Some imperial units are still used in the UK today.

improper fraction: a fraction where the numerator is greater than the denominator

$$\frac{8}{5}, \frac{20}{8}$$

inch: an imperial unit of measure used to measure length (1 inch is approximately 2.5 cm)

infinite: able to be continued without end

integer: a number without a fractional part; a whole number

..., –4, –3, –2, –1, 0, 1, 2, 3, 4, ...

interior angle: an angle created by moving one side of a shape onto an adjacent side, moving around the inside of the shape

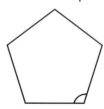

intersect: when two or more lines meet or cross

irregular polygon: a polygon that is not a regular polygon, its sides are not all the same length and its angles are not all the same

kilogram: a unit of measure used to measure mass (1 kilogram = 1000 grams)

length: a measure of how long something is; the distance between two points

lighter: less heavy; has less mass

line: a straight one-dimensional figure that has no thickness and extends infinitely in both directions

line of symmetry: a line through a shape that can be imagined or drawn that divides it into two parts that, when folded onto each other, will cover exactly the same shape and area

line segment: a straight line that connects two points

A ——————— B

litre: a unit of measure used to measure liquid (1 litre = 1000 millilitres)

mass: The amount of matter that something contains (usually measured in grams or kilograms). When we use a set of scales, the scales show the mass of the object.

maximum: the greatest amount or value possible

mental strategy: any method of working out the answer to a calculation in your head, without writing anything down

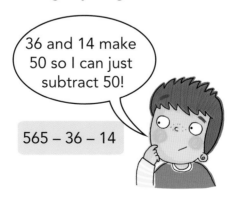

36 and 14 make 50 so I can just subtract 50!

565 – 36 – 14

metric: A system of measurement that uses metres, grams and litres. Metric measurement is based on multiples of 10, 100, 1000 and so on.

mile: an imperial unit of measure used to measure length (1 mile is approximately 1.6 kilometres)

millilitre: a unit of measurement used to measure liquid (1 millilitre = 0.001 litres)

—— 5 millilitres

minuend: the number being subtracted from: minuend – subtrahend = difference

minuend → 39 – 27 = 12

minus: subtract

mixed decimal: a decimal number with a whole number part that is not zero

1.7, 100.73564, 29.0051

mixed number: a number with a fractional part as well as a whole number part, written as a numeral and fraction together

$14\frac{1}{3}$

multiple: a number that may be divided by another a certain number of times without a remainder, for example some multiples of 5 are 5, 10, 15, 20, 25, 30, 35

multiply/multiplication: combining equal quantities; the symbol for multiply is ×

Glossary

multiplier: the second number in a multiplication; the number by which another number is to be multiplied

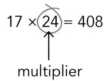

multiplier

natural number: positive numbers from 1; also called 'counting numbers'

1, 2, 3, 4, 5, 6 …

non-polygon: A shape that does not meet the requirements to be a polygon. (In order to be a polygon, a shape must have no curved lines, must be a closed shape and be two-dimensional.)

non-unit fraction: a fraction in which the numerator is not 1

$$\frac{3}{4} \qquad \frac{7}{8}$$

number line: A drawing of a line with positive and negative numbers in order, equally spaced and increasing in number from left to right from the origin to help counting. The line has arrows at both ends to indicate that it continues.

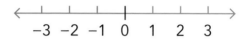

number sentence: a mathematical sentence written with numerals and symbols

$$583 + 73 = 656$$

numerator: the number of parts of an object, quantity or number that have been selected

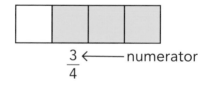

$\frac{3}{4}$ ⟵ numerator

obtuse angle: An angle that is bigger than a right angle but smaller than a straight angle. It is greater than 90 degrees but less than 180 degrees.

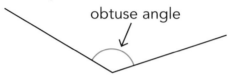

obtuse angle

octagon: a 2-D shape with 8 straight sides and 8 vertices

operation: a mathematical process; adding, subtracting, multiplying and dividing are all operations

order of operations: the order that operations should be completed in a calculation that has more than one step

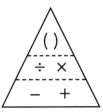

partition: split a number into parts that have the same total value

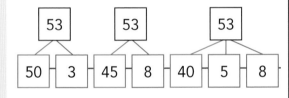

pentagon: a 2-D shape with 5 straight sides and 5 vertices

perimeter: the distance around the edge of a 2-D shape

perimeter = 30 cm

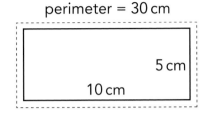

pint: an imperial unit of measure used to measure liquid (1 pint is approximately 560 ml)

place value: the value of a digit indicated by its position within the string of digits that represent a number

placeholder: a zero in a number that has no value itself but changes the value of other digits, for example 406

plus: add

point: an exact location in space

polygon: a 2-D shape with 3 or more straight sides

positive integer: any number that is 0 or greater and does not have a fractional part

0, 1, 2, 3, 4, …

pound: an imperial unit of measure used to measure mass (1 pound is approximately 450 g)

prime factor: a factor that is a prime number, for example the prime factors of 10 are 2 and 5

prime number: a number that is divisible only by itself and 1, for example 2, 3, 5, 7, 11

prism: a 3-D shape with two identical ends and flat faces

product: the answer when two or more numbers are multiplied together

35 × 4 × 2 = 280 ← product

Glossary

proper fraction: a fraction where the numerator is less than the denominator

protractor: an instrument used to measure angles

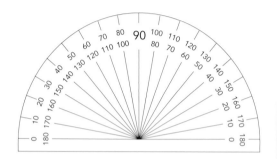

pure decimal: a decimal number that has a whole number part that is zero

0.7, 0.73564, 0.0051

pyramid: a 3-D shape with a polygon base and triangular faces that meet at the top (apex)

quadrilateral: a 2-D shape with 4 straight sides and 4 vertices

quotient: the result when dividing one number by another

In 5400 ÷ 30 = 180, the quotient is 180

radius: the distance from any point on the circumference of a circle to the centre

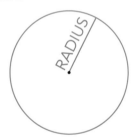

rectangle: A 2-D shape with 4 straight sides and 4 vertices. Opposite sides are equal in length. All angles are right angles.

rectangular: a shape with the properties of a rectangle

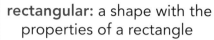

reflex angle: an angle that measures more than 180 degrees and less than 360 degrees

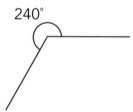
240°

regular polygon: a polygon that is equiangular (all angles are equal in measure) and equilateral (all sides have the same length)

related denominator: when one denominator is a multiple or factor of another denominator

remainder: the amount that is left over after a division when another equal group cannot be made

repeated addition: adding the same value repeatedly

$$5 + 5 + 5 + 5 = 20$$

Roman numeral: number written with symbols from the Roman times

1	5	10	50	100	500	1000
I	V	X	L	C	D	M

4	6	9	11	40	60	90	110	400	600	900
IV	VI	IX	XI	XL	LX	XC	CX	CD	DC	CM

112	340	400	504	600	799	949
CXII	CCCXL	CD	DIV	DC	DCCXCIX	CMXLIX

rounding (also called rounding off): approximating a number by expressing it to the nearest whole ten, hundred, thousand or other unit

rounding down: approximating a number by expressing it to the previous ten, hundred, thousand (or other unit)

rounding up: approximating a number by expressing it to the next ten, hundred, thousand (or other unit)

row: things positioned side by side or horizontally

ruler: an instrument used to measure length

second: a very short period of time; there are 60 seconds in a minute

side: the line joining two vertices of a 2-D shape

side length: the length of a side

simplify/simplified: Laws and rules about operations can often be used to make a calculation easier. This is called simplifying. For example, to change a fraction to an equivalent form with a smaller numerator and denominator.

single-digit number: A number with only one digit. These are the numbers less than 10.

solution: the answer to a calculation or problem

speed: how fast something travels over a given distance

43

sphere: A 3-D shape that is perfectly round like a ball. The distance from any point on its surface, through the centre to an opposite point, is always the same.

square: A 2-D shape with 4 straight sides and 4 vertices. All sides are the same length. All angles are right angles.

square centimetre (cm²): The area of a square that has length 1 cm and width 1 cm. Square centimetres are a unit of measure used to measure area.

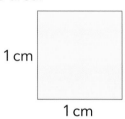

1 cm

1 cm

square metre (m²): The area of a square that has length 1 m and width 1 m. Square metres are a unit of measure used to measure area.

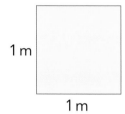

1 m

1 m

square number: the product of a number multiplied by itself

straight angle: an angle of 180 degrees

180°

subtract/subtraction: to find the difference between two numbers; the symbol for subtract is −

subtrahend: the number (27) being subtracted from the minuend; minuend − subtrahend = difference

subtrahend

39 − = 12

sum: the total of two or more numbers in an addition calculation

145 + 342 = 487

sum

symmetry: a shape has symmetry if it can be imagined as folded so that its 2 parts fit exactly onto each other, covering the same shape and area

times: another word for multiplied by, meaning 'lots of'

total: another word for 'sum'; the result of addition; the whole, all

$$\begin{array}{r} 2 \\ +\ 2 \\ \hline 4 \end{array}$$ ← total

triangle: a 2-D shape with 3 straight sides and 3 angles

unit fraction: a fraction in which the numerator is 1

unit of measure(ment): Standard amounts that are used to measure distance, area, capacity, mass, weight, time and so on. For example, some units of measure used to measure distance are millimetres, centimetres, metres and kilometres.

---

vertex/vertices: a point where two or more straight lines meet

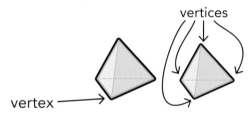

vertices

vertex

---

vertical: In an up and down direction. A table leg is vertical.

---

vertically opposite angles: angles opposite each other when two lines cross

---

volume: The amount of 3-D space occupied by an object. This includes solid and hollow objects. We calculate volume by multiplying the 3 dimensions together (length × width × height/depth).

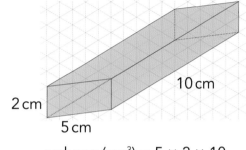

10 cm

2 cm

5 cm

volume (cm^3) = 5 × 2 × 10

---

weight: The force of gravity pulling on an object. For example, when astronaut Tim Peake was in space he was weightless because there is zero gravity – but his mass was still 70 kg.

---

whole number: a number without a fractional part, including a zero

0, 1, 2, 3, ...

---

width: a measure of how wide something is; the distance between two points

---

word calculation problem: a problem where words are used to describe a situation that involves numbers

---

zero: The number before 1 in our number system. Zero has no value.

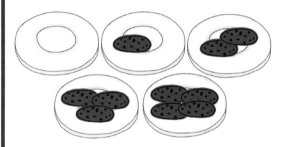

1	2	3	4	5	6	7	8	9	10
11	12	13	14	15	16	17	18	19	20
21	22	23	24	25	26	27	28	29	30
31	32	33	34	35	36	37	38	39	40
41	42	43	44	45	46	47	48	49	50
51	52	53	54	55	56	57	58	59	60
61	62	63	64	65	66	67	68	69	70
71	72	73	74	75	76	77	78	79	80
81	82	83	84	85	86	87	88	89	90
91	92	93	94	95	96	97	98	99	100

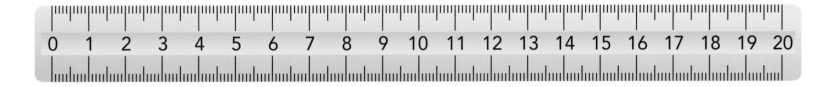

1	5	10	50	100
I	V	X	L	C

4	6	9	11	40	60	90
IV	VI	IX	XI	XL	LX	XC

18	34	59	92
XVIII	XXXIV	LIX	XCII